T0275657

Theoretical and Experimental Methods for Defending Against DDoS Attacks

Theoretical and Experimental Methods for Defending Against DDoS Attacks

Mohammad Reza Khalifeh Soltanian
Iraj Sadegh Amiri
University of Malaya, Kuala Lumpur, Malaysia

Matthew Neely, Technical Editor

ELSEVIER

AMSTERDAM • BOSTON • HEIDELBERG
LONDON • NEW YORK • OXFORD
PARIS • SAN DIEGO • SAN FRANCISCO
SINGAPORE • SYDNEY • TOKYO
Syngress is an imprint of Elsevier

SYNGRESS.

Syngress is an imprint of Elsevier
225 Wyman Street, Waltham, MA 02451, USA

ISBN: 978-0-12-805391-1

British Library Cataloguing-in-Publication Data
A catalogue record for this book is available from the British Library

Library of Congress Cataloging-in-Publication Data
A catalog record for this book is available from the Library of Congress

For information on all Syngress publications
visit our website at store.elsevier.com/Syngress

Working together
to grow libraries in
developing countries

www.elsevier.com • www.bookaid.org

CONTENTS

LIST OF FIGURES

LIST OF TABLES

In computer networking, distributed denial of service (DDoS) attack is one of the harmful attacks. It utilizes legal requests from hundreds or thousands of Internet users to specific targets to deplete both targets' bandwidth and resource. In fact, DDoS attack is an attempt to make a machine or network resource unavailable to its intended users. In this book, we have developed algorithm to detect and mitigate DDoS attacks theoretically and experimentally. The proposed algorithm has been successfully tested in several Internet service providers (ISP) networks. This algorithm also identifies the source of attacks even behind network address translation (NAT) so that routers or gateways communicate and collaborate to each other in order to identify suspicious traffic. This algorithm provides a new detection method to increase the attacks detection accuracy, which samples the current incoming traffic and CPU usage of the destination target of attack and calculates the difference from average. It then compares it to the predefined parameter β to distinguish whether or not the attacks happened. The threshold β measures the magnitude of traffic surge over the average traffic value or the magnitude of CPU surge over the average CPU value. If the firewall senses any attacks, it will send a request to any source edge routers to collaborate with them in order to detect and mitigate attacks. The proposed algorithm has been written based on cryptographic concepts such as birthday attacks to estimate the rate of attacks generated and passed along the routers. Consequently, attackers would be identified and prohibited from sending spam traffic to the server that can cause DDoS attacks.

CHAPTER 1

Introduction

1.1 DDoS ATTACKS

Denial of Service (DoS) attacks are now one of the biggest issues in the Internet. It refers to malicious attempts to prevent legitimate users from accessing requested resources by depleting bandwidth or depleting the resource itself. Distributed denial of service (DDoS) is a large-scale DoS attack, which is distributed in the Internet. Every computer, which has access to the Internet, can behave as an attacker. Typically, bandwidth depletion can be categorized into flood and amplification attack. Flood attacks can be done by generating ICMP packets or UDP packets in which it can utilize stationary or random variable port. Smurf and Fraggle attacks are used for amplification attack (Specht & Lee, 2004). DDoS *Smurf attack* is an example of an amplification attack where the attacker sends packets to a network amplifier with the return address spoofed to the victim's IP address (Specht and Lee, 2004).

Network amplifier is defined as a system capable of supporting broadcast addressing. Malformed packet and protocol exploit attacks are the two most important ways to deplete resources. Nowadays, each of them can cause different types of application layer attacks (Yu, Lu, Fang, & Li, 2009). Application layer DDoS attack is a DDoS attack that utilizes the communication protocol and sends out requests that they are indistinguishable from legitimate requests in the network layer. Moreover, most application layer protocols, which are built on TCP communicate with users using sessions, which includes one or many requests (Yu et al., 2009). HTTP1.0/1.1, FTP, and SOAP are examples of application layer protocols. An application layer DDoS attack is categorized in one or a combination of the following types (Ranjan, Uysal, Swaminathan, & Knightly, 2006):

1. Rate of session connection requests sent by session flooding attack are higher than legitimate users.

2. Request flooding attack sends sessions that include more requests than normal.
3. Asymmetric attack sends sessions with more high-workload requests.

Constrained bandwidth and resource power are the major servers' weakness points and to overcome these issues, we have to define specific threshold level for them, as they are guaranteed quality of service (QoS). False rejection rate (FRR) is the fraction of the rejected requests over the total number of requests from legitimate users. Similarly, false acceptance rate (FAR) can be defined as the fraction of the accepted requests over the total number of requests from illegal users or attackers. Although a good DDoS defense mechanism has to reduce both FAR and FRR, reducing FRR has more importance for the sake of user experience. Packet rate is used by most existing schemes as the metric to limit attackers (Li, Chang, & Chan, 2005). Using packet rate as a metric to limit damage from attackers is the idea that if the source of the attacks can be identified and traced-back incrementally hop-by-hop to the source (or as close as possible), then rate limiting can be used to limit the scope and damage of the attacks. One of the famous traced-back packets to find the source of attacks is IP traceback scheme proposed by (Park & Lee, 2001) in which packets are randomly marked for tracking the routes of the attack packets. Rate limiting reduces the rate of the packets allowed through to routers or gateways. It is possible that intelligent attackers can adjust their packet rate based on server's response to evade detection. On the contrary, visiting the clients' history logs are difficult to be modified, and the data used in trust evaluation are secured using cryptography. Hence, using trust as the evaluation criteria will be more reliable in application layer DDoS attacks. One of the major properties of our solution to identify and mitigate DDoS attacks, which is distinct from other solutions, is the manner in which routers and firewalls communicate to each other to reduce FRR and FAR as much possible as they can.

DDoS can take the major websites down for several hours at a time by (Dean & Stubblefield, 2001). The attackers are able to break into hundreds or thousands of computers or machines and install their own tools to abuse them. Then they utilize these "zombie" machines to launch the DDoS attack as shown in Figure 1.1. Some of the tools

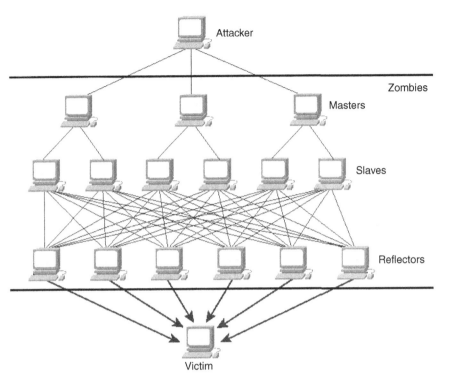

Fig. 1.1. A single zombie network performs a DDoS attack.

even use encrypted communications between the attacker and zombie machines. To make the detection of the zombie machines hard as much as possible, these tools counterfeit the source IP address on the traffic that they generate.

The tools work via brute force, which means random traffic is generated (may be with a political message) aimed at a specific machine. Most websites can bring down by generating gigabyte per second of traffic aimed at a single machine. For instance, the attacker could easily manage an attack such that the e-commerce sites remained available, but web surfers are unable to complete any purchases. Such an attack is possible by targeting the secure server that processes credit card payments instead. To make the clients' connection to such servers as secure as possible, the SSL/TLS protocol allows the client to request the server to perform a RSA decryption without having done any work first. One of the main disadvantages of RSA decryption is that it is an expensive operation. Only a few secure sites able to process 4000 RSA decryptions per second (Dean and Stubblefield, 2001). If we assume that a partial SSL handshake takes 200 bytes, then 800 KB/s is sufficient to paralyze an e-commerce site. Such a small amount of traffic is much easier to hide. Hence, this could be a major reason that DDoS attack can bring down even a very powerful server or router before taking any countermeasure. Up until now, there are lots of Internet service companies faced DDoS attack. These attacks not only make some problems to the Internet users, but also impose severe financial losses to the targeted Internet companies that their business depends on online availability such as Amazon, ebay, Zappos, etc. Moreover, a DoS attack can initiate other malicious activities such as worm or virus infection and steal confidential information. DoS attack can be classified into two categories, which are bandwidth and resource depletion. These categories are based on the inherent properties of the DoS attack, which either exhaust the resources of the server or attempt to deny the critical services. In Figure 1.2, the taxonomy of DoS or DDoS attack is illustrated.

1.2 MOTIVATION

Sources of information via the Internet and demands for access to the Internet servers such as file servers, web servers, and mail servers, increases exponentially. For instance, some malicious users known as spammers or spam DoS attackers abuse a popular server by sending spam request

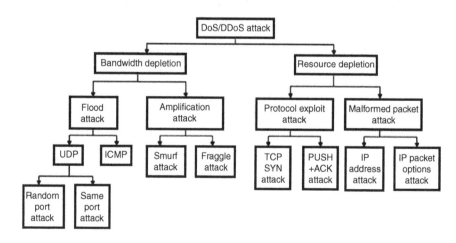

Fig. 1.2. *Taxonomy of DoS and DDoS.* From Mirkovic and Reiher (2004).

to overwhelm the server. Increasingly, spam requests are sent via zombie networks that work like a DDoS attack. Therefore, operations of the Internet service providers and other Internet companies who have partnership or collaboration with those providers will be affected and experience service disruption. As a result, it would be crucial to mitigate the spam DDoS attack by having a secure deployable protection scheme applied on routers and server in the network.

1.3 OBJECTIVE

The objective of this project is to propose a practical algorithm to allow routers to communicate and collaborate over the networks to detect and distinguish DDoS attacks. This algorithm allows the detection of DDoS attacks on the servers as well as identify and block the attacks.

1.4 ORGANIZATION

The remaining parts of this report are organized into four chapters.

In Chapter 2, a survey and the existing methods to identify and mitigate DDoS are presented. All related works are examined and discussed.

In Chapter 3, some preliminaries are covered, which include the introduction of Mikrotik routers as infrastructural routers, concept of birthday attacks, and birthday paradox. Also, we focus on the design and implementation of a comprehensive algorithm to identify and block source attacker. The workflow of the implementation is explained. Several real test cases, modeled to examine the efficiency of our theoretical and experimental implementation in defeating DDoS will be presented in the following chapter.

In Chapter 4, we present the results and discussions of our theoretical and experimental implementation. We also focus on obtaining an optimum parameter for router setting and make trade-offs to improve efficiency of mitigating DDoS attacks.

In Chapter 5, we conclude our achievement in this project based on whatever we found in our implementation. Last but not the least; we make recommendations for future work to further improve the algorithms.

CHAPTER 2

Related Works

2.1 GENERAL OVERVIEW AND DEFINITIONS

Numerous attempts like the followings have been made as DDoS defense and response mechanism that includes packet filtering (Kim, Lau, Chuah, & Chao, 2004), IP traceback (Aljifri, 2003; Bellovin, Schiller, & Kaufman, 2003; Houle et al., 2001), flood pushback (Ioannidis & Bellovin, 2002), and client puzzle theory (Fraser, Kelly, Raines, Baldwin, & Mullins, 2007).

The capability to recognize the addresses of the true sources of the packets causing a DDoS is made available for the victim's network administrators by IP traceback methods that also comprise the introductory phase of recognition of the identity of any invader controlling a number of compromised machines in the DDoS. To enable victims for finding the attackers, there is continuous study in IP traceback, identifying the real source of a spoofed packet. In the following, there are some problems with methods of IP traceback.

A slow complex process of logging into every router on the path may be regarded as one of manual ways. A number of costly routers might have processing abilities with internal software offering some stage of automatic IP traceback. A vigorous attack for the period of the testing of upstream links is required by link testing up to the time, when the source is located. Internet service provider (ISP) cooperation and management overhead are necessary for input debugging. Numerous layers, usually comprising of blameless victims, exist in a DDoS network; the instigator may not even be active in the time of occurring the attack. Limiting the rate of ICMP and/ or SYN packets, verifying for correct reverse paths, using Ingress/Egress filtering (checking out packets, having valid inside source address, and having outside source destination), and preventing packet delivery from unidentified hosts may be conducted by firewalls. Sending information to other sources or saving information on the packet destination for future use, and marking them with additional information are regarded as router solutions through software enhancements.

Research in attack traffic filtering/packet filtering can be categorized into three areas according to the protection point:

2.1.1 Source Initiated

Source sites are liable for assurance that there are attack-free outgoing packets. Examples comprise network ingress filters (Ferguson & Senie, 2000), disabling ICMP, or eliminating unexploited services to avoid computers from getting attack agents, or filtering odd traffic from the source (Mirkovic, Prier, & Reiher, 2002). The feasibility of these approaches, however, pivots on voluntary collaboration among a major number of ingress network administrators Internet-wide, making these methods somewhat unreasonable given the extent and the Internet uncontrollability.

2.1.2 Path-Based

Only the packets pursuing the correct paths are permitted in this approach (Kuzmanovic & Knightly, 2003). Every packet with an incorrect source IP for a particular router port is assumed as a spoofed packet and is dropped. This eliminates up to 88% of the spoofed packets (Park & Lee, 2001). In another approach (Jin, Wang, & Shin, 2003) for a source IP, if there is wrong number of travelled hops, the packet is dropped, eradicating up to 90% of the spoofed packets. These approaches are believed practical, but there is a somewhat high possibility of false negatives, ie, falsely accepting attack packets. It seems that none of these approaches work when packets use unspoofed addresses that are an emerging trend.

2.1.3 Victim-Initiated

Countermeasures to decrease received traffic can be started by the victim. For instance, in the pushback scheme (Ioannidis and Bellovin, 2002), the victim initiates decreasing excessive incoming traffic and asks for the upstream routers to conduct rate reduction, too. There exist other approaches according to packet marking (Kim, Jo, & Merat, 2003a; Xu & Guérin, 2005), an overlay network (Keromytis,

Misra, & Rubenstein, 2004), statistical processing (Kim et al., 2004; Li et al., 2005), TCP flow filtering (Kim, Jo, Chao, & Merat, 2003b; Yaar & Song, 2004), etc. Though victim-initiated protections are more favorable, some approaches are too costly to apply or need changes in Internet protocols.

On the other hand, traffic filtering method is costly, and confirmed to be susceptible in current attacks. Furthermore, ISPs are dependent on blocking of DDoS attacks and manual detection. If there is an attack, a subject-matter specialist executes an offline fine grain traffic analysis to recognize and distinguish the attack packets. Then, new filtering regulations on access control lists are created and fitted by hand on the routers. However, poor response time and fails to protect the victim are caused by the requirement for human interference before harsh harms are appreciated. In addition, the clarity of current rule-based filtering is too restricted, because it needs a clear specification of all kinds of packets to be removed.

As shown in Figures 2.1 and 2.2, an overlay network is a computer network made on the top of another network. In the overlay, nodes can be suggested of as being attached by logical or practical links, so that each of them matches to a lane, possibly by many physical links, in the original network. For instance, distributed systems like peer-to-peer networks, client-server applications, and cloud computing are overlay

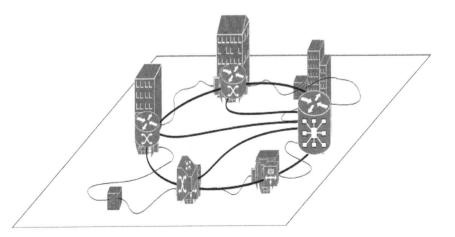

Fig. 2.1. A sample overlay network.

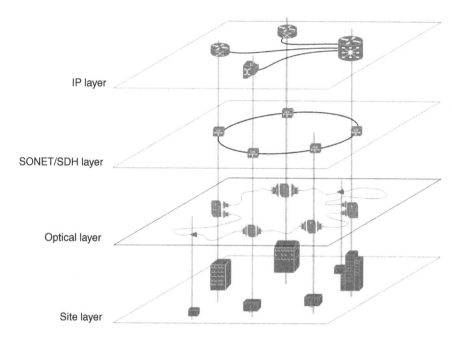

Fig. 2.2. Logical layers of overlay network. From Wikipedia.

networks, as their nodes run on top of the Internet. The Internet was basically made as an overlay upon the telephone network, while through the advent of VoIP, today there is the telephone network changing into an overlay network made on top of the Internet more than before.

Overlay networks are applied in telecommunication networks due to the accessibility of digital circuit switching tools and optical fibres. Telecommunications and IP networks, joined to promise the broader Internet, are all overlaid with at least an IP or circuit layers of the public switched telephone network, an optical layer and a transport layer. At first enterprise private networks were overlaid on telecommunication networks like frame relay and asynchronous transfer mode packet switching infrastructures. However, relocation from these infrastructures to IP based MPLS networks and virtual private networks began from 2001 to 2002 (AT&T History of Network Transmission).

In order to show Pushback, think about the network in Figure 2.3 in which the server D is under attack, and the routers R_n are the last few routers through which traffic get to the server D. Links by which attack

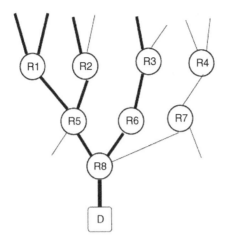

Fig. 2.3. A DDoS attack in progress. From Ioannidis and Bellovin (2002, p. 2).

traffic is running are illustrated by the thick lines, and the links with no bad traffic are shown through thin lines. Like the internal part of the network effectively conditioned, only the last link is actually congested. Hardly any nonattack traffic would be getting to the target in the lack of any particular measures. Some nonattack traffic is running by the links between R2–R5, R3–R6, R5–R8, R6–R8, and from R8 to D, but majority of it is dropped because of the R8–D congestion. *Bad* packets include which are transmitted by the attackers. An *attack signature*, which strived to identify characterizes bad traffic and the *congestion signature*, which is the group of characteristics of the aggregate identified as causing problems compromise something that can be really recognized. *Poor* traffic contains packets that set the congestion signature, however, they are not actually part of an attack that are just unsuccessful enough to have the similar target, or some other characteristics that make them identified as belonging to the attack. *Good* traffic does not match the congestion signature, but shares links with the bad traffic and may thus suffer.

Based on the figure mentioned earlier, the traffic entering R4 may be divided, some good like the part exiting R7 that is not belonging to R8 and some poor, as belonging to *D*. From the links mentioned earlier, there are some good traffic going to R5, and exiting from the lower left link, however, based on how congested the links R1–R5 and

R2–R5 are, it may suffer. Further links include a combination of good and poor traffic. Then, it can do nothing to permit more good traffic initiating from the graph left side to reach *D,* regardless of how smart filters R8 could employ. If more good traffic would run in via R7, all it can perform is specially drop the traffic from R5 and R6. By pushback, R8 transmits messages to R5 and R6 notifying *them* to rate-limit traffic for *D.* Though the links downstream from R5 and R6 are not congested, packets are going to be dropped anyway when they arrive at R8, so perhaps they are dropped at R5 and R6 as well. These two routers, in sequence, broadcast the demand up to R1, R2, and R3, notifying *them* to rate-limit the bad traffic, permitting some "poor", and more "good" traffic, to run through.

According to extending a scheme of measured signature generation to incorporate anomaly detection system (ADS) with Snort by extracting signatures from anomalies noticed, more sophisticated intrusion detection systems (Hwang, Cai, Chen, & Qin, 2007; Ning, Jajodia, & Wang, 2001) have been recently suggested. In addition, DDoS defense schemes (Chen & Hwang, 2006a,b; Kandula, Katabi, Jacob, & Berger, 2005; Moore, Voelker, & Savage, 2001; Walfish, Vutukuru, Balakrishnan, Karger, & Shenker, 2006) utilizing *Kill-Bots, Backscatter analysis,* and experimental evaluation of *speak-up*, recently there existed a defense against *application-level* DDoS.

In order to protect Web servers against DDoS attacks, *Kill-Bots* is a kernel extension that pretend as flash crowds. By providing authentication, Kill-Bots uses graphical tests that are different from other systems using graphical tests. First, to recognize the IP addresses that disregard the test, Kill-Bots uses an intermediate stage, and determinedly attack the server with demands despite frequent breakdowns at solving the tests. Since their intention is to obstruct the server, these machines are bots. On one occasion these machines are identified, Kill-Bots turns the graphical tests off, blocks their requests, and permits access to valid users, who are incapable or reluctant to resolve graphical tests. Second, without permitting unauthenticated clients access to sockets, TCBs, and worker processes, Kill-Bots launches a test and controls the client's reply. The expected computing base (TCB) of a computer system is regarded as some of all hardware, firmware, and/or software parts that are serious to its security, in the sense that bugs or susceptibilities that occur in the TCB might endanger the security features of the whole system. Contrastingly, sections of a computer system out of the TCB must not be capable to misbehave in a way that if there would be any leak more than what is granted to them based on the security policy. Thus, Kill-Bots are expected to protect the authentication mechanism from being DDoS. Third, Kill-Bots joins authentication with admittance control. A few restrictions are taken into account as the followings. First, Kill-Bots complicatedly deals with Web Proxies and NATs that multiplex a single IP address among several users. In case, all clients behind the proxy are legitimate users, then sharing the IP address has no impact. Contrastingly, when a zombie shares the proxy IP with legitimate clients and uses the proxy to mount an attack on the Web server, all subsequent requests from the proxy IP address may be blocked by Kill-Bots. Second, Kill-Bots has a few allocated parameters based on the experience. Third, Kill-Bots supposes that the first data packet of the TCP link will include the *GET* and *Cookie* lines of the HTTP demand.

In an attack, *Backscatter analysis* supposed a per-packet random source address, consistent delivery, and one reply generated for each packet. It states the possibility of a given host on the Internet obtaining at least one unsolicited reply from the victim is $m/2^{32}$ in an attack of m packets. Likewise, if one checks n distinct IP addresses, then the hope of occurring an attack would be:

$$E(x) = \frac{nm}{2^{32}}$$

Viewing a large enough address variety can "sample" such denial-of-service activity on the Internet. In these samples are information about the kind of attack, the victim's identity, and a timestamp from which can estimate attack period. Furthermore, considering these suppositions, can lead to use the average arrival rate of unsolicited responses directed at the monitored address range to approximate the real speed of the attack directed at the victim, as follows:

$$R \geq R' \frac{2^{32}}{n}$$

Supposing that R' is the backscatter calculated average inter-arrival rate from the victim and R is the extrapolated attack rate in packet-per-second, there are three main suppositions that underlie *backscatter analysis*:

- *Address uniformity.* Attackers spoof source addresses randomly.
- *Reliable delivery.* Attack traffic is delivered consistently to the victim and backscatter is delivered consistently to the monitor.
- *Backscatter hypothesis.* Unsolicited packets viewed by the monitor signify backscatter.

Amid these assumptions, key includes the random collection of the source address. There exist some explanations why this supposition may not be legitimate. First, *ingress filtering* (Fullmer & Romig, 2000) is employed by some ISPs on their routers to drop packets with source IP addresses outside the range of a customer's network. Consequently, an attacker's source address range might not comprise any of our intended addresses and we will undervalue the whole number of attacks.

"Reflector attacks" present a next problem for source address consistency. In this condition, by sending a packet spoofed with the victim's source address to a third party, an attacker "launders" the attack. And by sending a response back toward the victim, the third party responds. Third parties may further intensify the attack, when the packets to the third party are addressed applying a broadcast address (as with the popular smurf or fraggle attacks). The significant matter with reflector attacks is that the source address is particularly chosen. IP address is unable to view the attack, if it occurs in the range, which is monitored as a reflector.

One more restriction is caused by *backscatter analysis* supposition that packets are delivered consistently and that every packet produces a response. Probably throughout a great attack, packets from the attacker may be queued and dropped. Firewall or intrusion detection software may filter or may rate limit those packets that *do* arrive. Furthermore, some forms of attack traffic (eg, TCP RST messages) do not normally extract a reply. Last, the responses may be queued and dropped themselves along the path back to the monitored address range. As with random allocation supposition, these restrictions will lead to *underestimate* the number of attacks and their rate. However, they might also bias the classification of victims. The final restriction of *backscatter analysis* technique is that it supposed unsolicited responses symbolize backscatter from an attack. There is freedom for any server on the Internet to transmit unsolicited packets to the observed addresses, and there may be misinterpretation of these packets as backscatter from an attack.

Speak-up, that is considered a defense against *application-level* DDoS, in which attackers cripple a server by sending legitimate-looking demands that use computational resources (eg, CPU cycles, disk). Through speak-up, all clients and resources permitting are encouraged by a victimized server *to send higher quantities of traffic repeatedly*. It assumes that most of the attackers' upload bandwidth is already being used, so that it cannot respond to the server. However, good clients have extra upload bandwidth, so that they can respond to the server through severely higher volumes of traffic. The planned result of this traffic inflation is that the bad clients are crowded out by the good ones through which there would be capturing much larger fraction of the server's resources than before. *Speak-up* makes the server use resources on a group of clients in rough quantity to their collective upload bandwidths.

There have been many attempts by some of the researchers to defeat repeated DDoS attacks (Hussain, Heidemann, & Papadopoulos, 2006). In their endeavour, an approach has been performed to fingerprint and recognize the frequent attack scenarios. Such fingerprints are assumed to not only help in the prosecution of criminal and civil trial of attackers, but also assist in validating and focusing on response measures. As packet contents may be easily controlled based on the spectral description of the attack flow, which are hard to falsify.

Some others apply trust-negotiation (Ryutov, Zhou, Neuman, Leithead, & Seamons, 2005) methods to establish trust, overlay networks (Wang, Chellappan, Boyer, & Xuan). Trust negotiation includes a method that offers an open verification and access-control surroundings for such transactions, but it is vulnerable to malicious attacks that result in denial of service or sensitive information leakage.

DDoS-resilient scheduling (Ranjan, Swaminathan, Uysal, & Knightly, 2006), D-WARD (Mirkovic & Reiher, 2005) and multilevel tree for online packet statistics (MULTOPS) (Gil & Poletto, 2001) were suggested for filtering and rate limiting on the flows supposed at the source-end. Often, security managers tend to prefer to focus more on defending their own networks, hence they selected home detection methods (Carl,

Kesidis, Brooks, & Rai, 2006). COSSACK (Papadopoulos, Lindell, Mehringer, Hussain, & Govindan, 2003) and DefCOM (Mirkovic and Reiher, 2005) organize detectors at the victim side and transmit an alert to the filter or to the rate limiter that is placed at the side of the source. Chen and Song (2005) suggested a perimeter-based scheme for Internet service providers to enable a service to defence DDoS attacks for their customers. They suggested a plan according to edge routers to realize the sources of the flood off attack traffics.

Change-point detection theory is used by most of the researchers to distinguish any unusual traffic distributed through the Internet due to DDoS attacks (Blazek et al., 2001; Chen & Hwang, 2006a,b; Peng, Leckie, & Ramamohanarao, 2003; Wang, Zhang, & Shin, 2004). Because of the lack of precise statistics to explain about the prechange and postchange traffic distributions, there would be a development of a non-parametric cumulative sum (CUSUM) scheme for its low-computational intricacy (Blazek et al., 2001). The short-term behavior shifting from a long-term one is monitored by the scheme monitors. If the cumulative difference arrives at a specific threshold, there would be an attack alert. A central DDoS defense scheme was proposed by Wang et al. (2004) to check the change points of the gateway level. A similar approach was taken by Peng et al. (2003) in the source IP addresses monitoring.

Here, we are going to review two remarkable efforts, which are used more recently and they are client puzzle theory and collaboration of detection of DDoS attacks over multiple networks.

2.2 CLIENT PUZZLE THEORY

Much of the presented solution to overcome DDoS attacks includes enforcing servers to produce a puzzle, when the servers feel under attacks and clients must resolve these puzzles and send it again to servers as shown in Figure 2.4. Verifying the solution by servers is the next step. Generally, framework applied to explain about any certain client puzzle's scheme includes main step proposed by Jeckmans (2009) and they are: *Setup*, *PuzzleGen*, *PuzzleSol*, and *PuzzleVer*. In general, they include the algorithms applied to open the requirements for puzzles formation, produce the client puzzles, solve the client puzzles, and verify the solution of the client puzzle for the client puzzle system correspondingly.

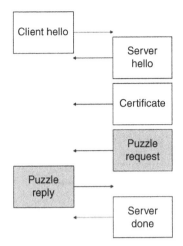

Fig. 2.4. General client puzzle handshaking. From Dean and Stubblefield (2001).

Setup(k)

It is run by the server S. It takes a security parameter k as input and output *params* as the system parameter, which would be an implicit input to the following algorithms. This system parameter is varied across different puzzle schemes.

PuzzleGen(mk, req)

It is run by the server S. It takes a server secret *mk* and additional request information *req* received from the client C as input and outputs a puzzle *puz* and additional information *data*, which required by the server for verification purpose. The puzzle *puz* is sent to client C.

PuzzleSol(puz)

It is run by the client C. It takes a puzzle *puz* that received from the server S and outputs a puzzle solution *sol*. The puzzle solution *sol* is sent to the server.

PuzzleVer(data,mk, sol)

It is run by the server S. It takes the puzzle information *data*, the server secret *mk*, and a puzzle solution *sol* received from the client C as input. It outputs 1 if *sol* is correct or 0 otherwise.

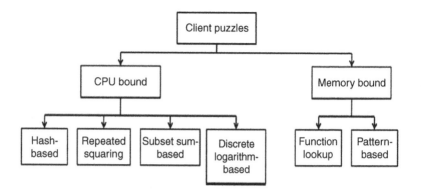

Fig. 2.5. Classification of client puzzles schemes. From Mirkovic and Reiher (2004).

Based on Figure 2.5, two main groups, CPU-bound puzzles and memory-bound puzzles can be regarded as categorization of the existing client puzzles schemes. While memory-bound puzzles need some memory lookups to be performed to solve them, CPU-bound puzzles are the puzzles that require CPU cycles to be performed when solving them. Therefore, the solving speed would deeply rely on the processor and machine memory speed.

Merkle (1978) suggested the first notion of cryptographic puzzles incorporated into network authentication protocol. Based on Merkle's idea, there are several client puzzles schemes, which are applied to frustrate DoS resource depletion types of attacks (Karame & Čapkun, 2010; Dean & Stubblefield, 2001; Feng & Kaiser, 2007a, b).

Several issues exist in performance of client puzzle, which have to be carefully weighted up as the following:

1. The price of puzzles' formation and answer confirmation used by the server has to be more costly than the price of solving the puzzle used by the client.
2. The complexity of puzzle ought to be modifiable.
3. There should be statelessness in the puzzle, so that the server may not need to keep any link state prior to the offering of puzzle key.
4. For the clients to solve the puzzle, only limited amount of time is given.

5. There should be infeasibility in precomputation attacks.
6. There is uniqueness in the puzzles, that is, by having the previous puzzles solved, it is not possible to solve new puzzles.
7. There must not be any flooding attack vulnerabilities in puzzle creation and issuing.
8. Susceptibility to any bypass issue must not exist in puzzle mechanism.
9. There must be a differentiation in puzzle issuing between legitimate users and malicious users, therefore fine users exhibit spiteful behavior in more complex puzzles.

The analysis of a client puzzles scheme can be based on four major factors:

1. *Server cost.* It consists of precomputation cost, construction cost, and verification cost. Generally, it determines the computational effort on the server's machine.
2. *Client cost.* It identifies the computational effort on the client's machine.
3. *Hardness granularity.* It measures the relationship between the workload needed to solve the puzzle and the puzzle difficulty. Fundamentally, there are three different types of hardness granularity; linear, polynomial, and exponential. The preferred granularity in puzzle scheme is the linear granularity while the worst case to deal with is the exponential granularity.
4. *Deterministic or nonparallelizability.* It represents the ability of the puzzle scheme to resist parallel computation attack. In other words, parallelizable puzzles can be distributed by coordinated adversaries to high performance computing machines, in order to perform parallel computation on the solutions of the puzzles.

Since the use of client puzzles as a mechanism to countermeasure DoS attack, most of the research has been focused on improving its security rather than developing a practical client puzzles system. The security improvement and deployment complexity of existing client puzzles schemes are not balanced. The limitations of existing client puzzles schemes can be summarized as follows:

1. Nonparallelizable is one of the important features a client puzzles scheme should have, in order to thwart the DoS attack efficiently.

It can be noticed that all the hash-based puzzles suffers from this shortcoming, no matter how much they have been improved for years.

2. Probabilistic behavior of a client puzzles scheme allows an unknown amount of solutions that can be obtained at the first try of solving it. The chances of such probabilistic occurrence will increases when the number of trial increases. This is not a desirable circumstance that should occur in a real application of client puzzles.

3. Low-differentiability difficulty setting of a client puzzles system unable to distinguish legitimate users from malicious users. This makes the system unfair to those legitimate users who pay the same cost as the malicious users. For a client puzzles system to defeat DoS, the pricing of the puzzles must be fair across different kind of users.

4. Complicated deployment causes the client puzzles systems, which are not preferred in real life application. Most of the secure client puzzles systems require the adoption of a specific client software or modification of a particular protocol in order to work. Such poor adaptability leads to inconveniency problems, such as compatibility and privacy protection issues. Only robust system that can survive in today advanced technology.

2.3 CPU-BOUND PUZZLES

CPU-bound puzzle is the largest class of client puzzle as compared to memory-bound puzzle. It requires the clients to commit their processor's resources when solving the puzzles. In the following sections, we will first describe the algorithms of the puzzles schemes and then analyze their performance.

2.3.1 Hash-Based Puzzle

Hash-based client puzzle schemes require the client to reverse a cryptographic hash function, in order to solve the puzzle. When utilizing hash function in client puzzle scheme, a part of the reverse is provided so that the client is able to solve the remaining part by brute force (Juels & Brainard, 1999).

2.3.1.1 Juels and Brainard's Hash Puzzle

In Juels and Brainard's approach (Juels and Brainard, 1999) to defend against connection depletion attacks, the cost of each connection request is a unique cryptographic problem called client puzzle. When no attack is taken place, the server accepts connection requests as usual. However, when the server observes an attack, each client that wishes to establish the connection must solve the client puzzle given.

2.3.1.2 Aura et al.'s Hash Puzzle

Aura et al. (2001) pointed out the weakness of Juels and Brainard's puzzle that neglects the DoS attacks against the authentication protocols. For the improvement on the efficiency of the client puzzles, they have reduced the length of the puzzle and its solution, minimized the cost of verification of the solution, and broadcasted the puzzles if the network allowed. To make the scheme more secure, signature is used to authenticate the communication between server and client.

2.3.1.3 Parallel Hash Puzzle

Parallel hash puzzle is originated from the single hash puzzle by Juels and Brainard (1999). In this scheme, a puzzle is divided into multiple smaller hash-based puzzles and a single client is asked to solve multiple hash-based puzzles in parallel. The purpose of doing this is to obtain linear puzzle hardness granularity by utilizing multiple hash puzzles in parallel where the client is responsible to find the missing bits of a pre-image of a hash function whose output is given.

2.3.1.4 Hinted Hash Puzzle

Hinted hash puzzle is proposed by Feng et al. (2005) to solve the puzzle hardness granularity issue. As, how hinted hash puzzle was named, a hint will be provided along with the puzzle to be solved by the client. The hint is used to achieve linear hardness granularity, which has the same purpose of multiple hash-based puzzles proposed by Juels and Brainard. By adjusting the accuracy of the hint, which is a single value that tells the client where the answer lies, the difficulty of the puzzle is adjusted.

2.3.1.5 Chained Hash Puzzle

A chained hash puzzle, proposed by Groza and Petrica (2006) consists of a number of hash-based puzzles, which may be solved only in a precise manner. In other words, the puzzles are chained together in a linear or random fashion making the puzzle solution dependent on each other based on how they are chained.

2.3.1.6 Targeted Hash Puzzle

A different kind of public work functions known as targeted cryptographic hash function reversal has been proposed by Feng and Kaiser (2007a, b). For a public work function to be practical, it must fulfill four important properties, which are fast issuing, fast verification, flexible adaptability, and precomputation and replay resistance.

2.3.2 Repeated Squaring Puzzle

The idea of time-release crypto was proposed by Rivest et al. (1996) to hide a message that cannot be revealed by anyone even the sender until, a predetermined amount of time has passed. This idea is then applied onto client puzzles called timelock puzzles or repeated squaring puzzles in which a precise amount of time is required to solve the puzzles. To solve a repeated squaring puzzle, the client has to perform modular squaring repeatedly and the number of repetition is determined in the puzzle.

2.3.3 Discrete Logarithm-Based Puzzle

Waters et al. (2004) proposed an idea of outsourcing the puzzle creation and distribution to a bastion, they use a new one-way function based on the discrete logarithm problem that can replace the role of hash function in their client puzzles.

2.3.4 Subset Sum-Based Puzzle

Another limitation of previous puzzle schemes is lack of parallel computation attack resistant, thus susceptible to DDoS attacks. Realizing the importance of nonparallelizability property in a puzzle, Tritilanunt et al. (2007) proposed subset sum puzzles. It was claimed by Tritilanunt et al. that subset sum puzzles is a new technique that offers simple construction and low cost verification as hash-based puzzles, it also has a nonparallelizable characteristic. Basically, it is just a subset sum problem that is used to be solved as a client puzzle. The fastest algorithm at the time of this writing to solve this problem is one of the lattice reduction algorithms called the LLL algorithm

(Lenstra et al., 1982). Another intention of requiring clients to implement the LLL algorithm in solving subset sum puzzles is its nonparallelizable property. Two different LLL lattice reduction techniques that can be used to solve a subset sum problem are backtracking or brute force searching and branch and bound method. Both have been used to make a performance comparison with LLL algorithm in (Tritilanunt, 2010).

2.3.5 Modified Time-Lock Puzzle

The traditional time-lock puzzle was first proposed by Rivest et al. (1996). The term time-lock is used by Rivest to describe the primer property of the puzzles in which a predetermined amount of time is required before the puzzles are available to be solved. It is also known as repeated squaring puzzles, as it was first employed in a client puzzle scheme, which a number of repeated squaring needs to be performed to solve it. There are some improvements that have been done on time-lock puzzles scheme by Feng and Kaiser (2010) to make it a more practical and secure scheme. The design of modified time-lock scheme separates the costs into puzzle verification cost and puzzle issuance cost. Basically, it is the combination of single hash puzzle and time-lock puzzle. The single hash puzzle will adopt in the puzzle issuer and the time-lock puzzle is responsible for the puzzle solver. This scheme is designed in this way because single hash puzzle offers linear hardness granularity and cheaper puzzle creation cost, which makes it the most suitable candidate as a puzzle issuer. As for time-lock puzzle, it is deterministic in nature and has parallelization resistance, which is the only scheme that satisfies the requirements as a puzzle solver. In summary, the modified time-lock puzzle has joined the strength and eliminated the shortcomings of single hash puzzle and time-lock puzzle.

2.4 SUMMARY OF CPU-BOUND PUZZLES

Cryptographic hard problems such as discrete logarithm problem, Diffie-Hellman problem, subset sum problem, and RSA problem have played a vital role in client puzzle scheme. Similarly, calculate the inversion of a one-way function such as cryptographic hash function, which is also considered as a mathematical hard problem, has been utilized to form the basis for most of the CPU-bound puzzles schemes. It can be observed that the first hash puzzles proposed by Juels and Brainard (1999) has some limitations, which then leads to a lot of research done to improve their scheme. All of these improvements are tailored, according to the properties of a good puzzle scheme.

First, the hardness granularity can be improved by distributing multiple client puzzles for the client to solve in parallel (Juels and Brainard, 1999). However, the best to granularity is by providing a range of possible solutions for the client to search in Feng et al. (2005) when the server's cost is taken into consideration. A stateful puzzle scheme might cause the server to be overloaded by the puzzle information. To make a puzzle scheme becomes stateless, the puzzle relevant information should not be stored in the server, and instead, they are sent to the client and have the client to send them back together with the solution. This must be done carefully without sending the information that could breach the security of the scheme. Even though most of the limitations in hash puzzle have been countered, the probabilistic nature of hash function causes it vulnerable to parallelization attack until today. Repeated squaring puzzle is one of the schemes that can resist parallel computation attack. Due to its high-storage cost and high-computational cost resulted from storing RSA modulus and calculating modular exponentiations respectively, it is considered impractical. Subset sum-based puzzle can utilize LLL algorithm's nonparallelizability to defend against parallelization attack. Nonetheless, subset sum-based puzzle suffers from being stateful and has a hardness granularity that is polynomial in nature. Last, modified time-lock puzzle has the best overall performance. It simply combines the hash-based puzzle scheme and repeated squaring puzzle scheme into one so that the advantages from both schemes are inherited, at the same time eliminating their disadvantages.

2.5 MEMORY-BOUND PUZZLES

Due to the great disparity in processing speed than in memory speed in today's computing hardware, memory-bound puzzles are introduced to make the puzzle solving less dependent on hardware. Primarily, the client is required to perform some look-ups in a precomputed database.

2.5.1 Function Look-Up Puzzle

Abadi et al. (2005) proposed a family of moderately hard memory bound functions that can be used to perform searching on a look-up table, which in turn increase the speed of solving a client puzzle. In Abadi's definition, a memory-bound function works on the ineffectiveness of the caches available in a machine. This can be done by having the memory-bound function behaves randomly, while accessing locations in a large region of the memory. In this situation, memory latency is used to measure the performance of a machine instead of memory throughput, which is less uniform. In short, a good memory-bound function will have a uniform cost across systems from high-end to low-end.

2.5.2 Pattern-Based Puzzle

A new memory bound puzzle construction based on heuristic search for sliding tile problem (Loyd & Gardner, 1959) using a look-up table called pattern database or heuristic table was proposed by Doshi et al. (2006). Basically, sliding tile problem is about finding a path to slide the tiles, which have been arranged in a specific pattern on a grid from a starting state to the goal state. The path used is not necessary to be the optimal path. In this situation, existing algorithms that were used to obtain the shortest path, they cannot be used always to find the solution. Optimal solutions to the 4×4 sliding tile problem can be acquired by using pattern databases introduced by Culberson and Schaeffer (1996). The main purpose behind incorporating pattern databases into the client puzzle scheme is that they consume more memory rather than processor resource, when search time is reduced desirably. Thus, puzzle-solving process is memory-bound.

2.6 SUMMARY OF MEMORY-BOUND PUZZLES

Existing memory-bound puzzles schemes do not eliminate the parallelization attack efficiently. All of them only offer polynomial hardness granularity. Due to their probabilistic behavior, the puzzle can be solved at first try. Their designs concern only the memory latency or memory speed by assuming an average memory size on clients' hardware. There is a possibility that some clients with huge memory size can store more pre-computed databases for solving the puzzle. So, attempt to slow down the clients by forcing them to renew their databases only work on those with smaller memory size.

2.7 COMPARISON OF EXISTING CLIENT PUZZLES SCHEMES

The abbreviations that describe the elements to be compared are presented in Table 2.1. The performance overview of existing client puzzles scheme is shown in Table 2.2. Both tables are adapted from Jeckmans (2009) with minor modifications made. Targeted hash puzzle scheme is added to the original Table 2.2. The puzzle cost is divided into puzzle creation cost and puzzle verification cost. The communication cost is approximated as the total cost of the puzzle scheme. Parallel computation resistance describes the ability of the scheme to defend against parallelization attack. Hardness granularity represents the relationship between the workloads needed to solve the puzzle and the puzzle difficulty with linear being better and exponential being worst. All schemes will have either deterministic nature or probabilistic nature. The comparison of existing client puzzles scheme can be summarized in the subsequent section. Most schemes are probabilistic in nature and vulnerable to the parallel computation attack. This indicates that these schemes will fail to defend against DoS and DDoS attack, which is launched by high-end machines.

Table 2.1 Abbreviations of Comparative Element

Header		Operation	
DN	Deterministic nature	H	Hash function
PC	Puzzle creation	M	(Modular) multiplication
PY	Puzzle verification	E	Modular exponentiation
PR	Parallel computation resistance	C	Checksum function
HG	Hardness granularity	F	One-way many-to-one function
LS	Long-term storage	Size	
SS	Short-term storage	k	Security parameter
CC	Communication cost	h	Hash value
		r	Modulus of RSA
Value			
l	Puzzle difficulty	s	Maximum puzzle difficulty
n	Number of puzzles	t	Maximum number of puzzles

Table 2.2 Comparison of Existing Client Puzzles Schemes

Client Puzzle Schemes	PC	PV	LS	SS	CC
Hash-based-Juel	$2H$	H	k	–	$3h+2k+2k^2$
Hash-based-Aura	–	H	–	$3k$	$3k$
Parallel hash	$2n*H$	$n*H$	k	–	$3n*h+2k+2k^2$
Hinted hash	$2H$	H	k	–	$3h+2k+2k^2$
Chained hash	$2n*H$	–	k	$n*h$	$3n*h$
Targeted hash	–	H	k	–	$h+k$
Discrete logarithm	E	–	k	k	$3k$
Repeated squaring	M	$2E$	–	$3r$	$3r$
Subset-sum	$H+1*M$	–	$s*h+k$	h	$3h$
Modified time-lock	$2H$	$2E$	–	$3r$	$3r$
Function look-up	$(1-1)F+C$	–	–	k	$3k$
Pattern-based	$1*C+H$	H	$(t+1)k$	–	$h+(2n+1+1)k$

Client Puzzle Schemes	PR	HG	DN
Hash-based-Juel	no	exponential	no
Hash-based-Aura	no	exponential	no
Parallel hash	no	polynomial	no
Hinted hash	no	linear	no
Chained hash	some	polynomial	no
Targeted hash	no	linear	no
Discrete logarithm	no	linear	no
Repeated squaring	yes	linear	yes
Subset-sum	yes	polynomial	yes
Modified time-lock	yes	linear	yes
Function look-up	some	polynomial	no
Pattern-based	some	polynomial	no

In order to prevent parallelization attack, the puzzle should be constructed in such a way that its workload cannot be distributed. Only the hash-based puzzles and modified time-lock puzzles schemes offer inexpensive server cost. We could observe that most schemes have cheap puzzle creation cost but expensive puzzle verification cost. Another observation is the improvement of hash-based puzzle in linearity. A hash-based puzzle can have a linear hardness granularity by providing some hints to the client to search for the solution. This has been done with the hinted hash puzzles. Most schemes require short-term storage and long-term storage space to keep track of some puzzle states. Puzzle scheme that requires short-term storage is a stateful scheme. As long as storage is needed in the scheme, the server's storage could be exhausted by attackers.

After the comparison among the existing client puzzles schemes, modified time-lock puzzle is found to be the only scheme that exhibits good overall performance. Therefore, it is chosen as the scheme to be implemented in this project.

Because of the application of client puzzles as a countermeasure against DoS attack, instead of developing a practical client puzzles system, most of the investigation has been centered on improving its security. There is any balance in the improvement of security and operation complexity of existing client puzzles schemes.

In general, client puzzle scheme cannot detect the attacker source and also imposes both server and client heavy unnecessary computation. This is the inherent weakness property of client puzzle.

2.8 COLLABORATION OF DETECTION OVER MULTIPLE NETWORKS

An algorithm was proposed by Chen, Hwang, and Ku (2007) in which a distributed method is proposed to identify DDoS flooding attacks at the traffic flow level. The defense system is appropriate for performance over the core networks operated by ISPs. Some traffic fluctuations are detectable at Internet routers or at the gateways of edge networks at the early stage of a DDoS attack. There is a development of a *distributed change-point detection* (DCD) architecture applying *change aggregation trees* (CAT). The chief idea includes detecting abrupt traffic changes over multiple network domains at the initial time.

A new mechanism, called CAT is applied by the algorithm that proposes a DCD architecture.

The flooding traffic that is sufficiently large to crash the victim machine through communication buffer overflow, disk exhaustion,

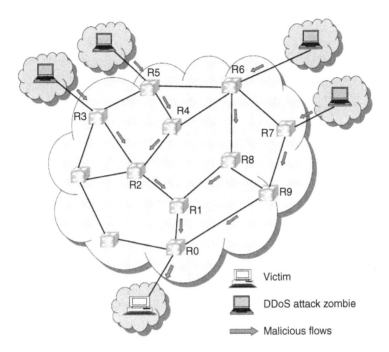

Fig. 2.6. *A large number of zombies generate traffic superflow by DDoS flooding attacks toward a common destination as victim host.* From Chen et al. (2007).

connection link saturation, and so forth. Figure 2.6 illustrates a flooding attack that is initiated from four zombies. The *attack-transit routers* (ATRs) identify the abnormal surge of traffic at their I/O ports. The attraction for the victim is by the end router *R0* in Figure 2.6. All the attack flows from the superflow homing toward the end router.

A superflow includes all packets destined for the similar network domain from all possible source Internet protocol (IP) addresses and uses a variety of protocols like transmission control protocol (TCP) or user datagram protocol (UDP), etc.

Briefly, Chen et al. (2007) developments may be classified in four technical features, as the quick vision and evidences of which are given in following segments:

1. *Traffic anomaly detection at the superflow level.* Monitoring Internet traffic at routers on individual flows is identified by a 5-tuple: source

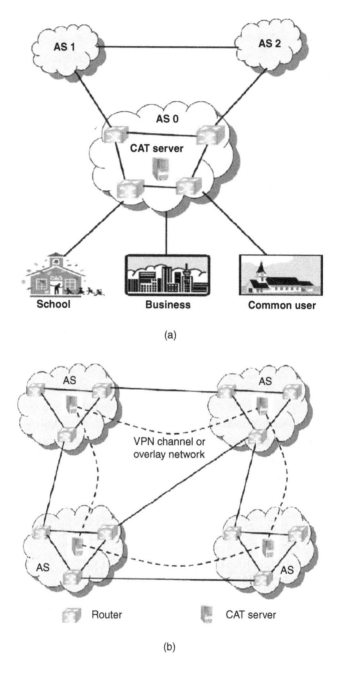

Fig. 2.7. *Distributed change detection of DDoS attacks over multiple AS domains. (a) Multidomain DDoS defense system. (b) Interdomain communication via VPN tunnels or an overlay network atop the CAT servers in four domains.* From Chen et al. (2007).

IP, destination IP, source port, destination port, protocol applied. The Superflow includes those traffic flows destined for the same network domain and exploits the same protocol. For DDoS defense in real-life Internet environments, this level of traffic monitoring and anomaly detection is more cost-effective.

2. *Distributed change-point detection.* The DCD scheme detects DDoS flooding attacks by monitoring the propagation of abrupt traffic changes inside the network. If a CAT tree is constructed sufficiently large and the tree size exceeds a preset threshold, an attack is declared.

3. *Hierarchical alerts and detection decision making.* The system accepts a hierarchical architecture at the domain and router levels. They suppose that their system simplifies the alert relationship, global detection processes and allows the DCD system to implement in ISP networks.

4. *Novelty of SIP.* They assume a trust-negotiating SIP to secure interserver communications. The SIP has removed some of the shortcomings of the existing IP security (IPsec) and application-layer multicasting protocols (Kent & Atkinson, 1998; Wang et al., 2006). SIP appeals for implementation on virtual private network (VPN) tunnels or over an overlay network built on top of all domain servers.

Through monitoring the propagation patterns of abrupt traffic changes at distributed network points, the DCD scheme is supposed to detect DDoS flooding attacks. An attack is asserted, if a sufficiently large CAT is constructed to exceed a preset threshold. The principles behind the DCD system are presented by this behavior.

The system architecture of the DCD scheme is illustrated in Figure 2.7. The system is organized over multiple AS domains. In each domain, there is a central CAT server. Traffic changes, aggregates suspicious alerts, checks flow propagation patterns, and merges CAT subtrees from collaborative servers into a global CAT is detected by the system. The root of the global CAT is at the victim end. Each tree node and each tree edge correspond to an ATR and to a link between the ATRs respectively. The abnormal surge of traffic at their I/O ports is detected by the *attack-transit routers* (ATRs).

CHAPTER 3

Problem Solving, Investigating Ideas, and Solutions

3.1 MIKROTIK ROUTERS

MikroTik is a Latvian company, which was established in 1995 to develop wireless and routers ISP systems. Nowadays, MikroTik provides software and hardware for the Internet connectivity in most of the countries all over the world. They created the RouterOS software system in 1997, regarding to their experience in using industry standard PC hardware and complete routing systems, which provides vast controls, flexibility, and stability for all kinds of routing and data interfaces.

Linux-based operating system known as MikroTik RouterOS is the main product of MikroTik. By installing the Mikrotik RouterOS on a personal computer or server computer, it turns the computer into a network router and enables implementing features such as virtual private network (VPN) server and client, firewall rules, wireless access point functions, bandwidth shaping, quality of service, and other commonly used features for interconnecting networks and routing. Moreover, this system is able to serve as a captive-portal-based hotspot system. The operating system is licensed to promote service levels, as far as each releasing provides more available RouterOS features. Winbox, a Windows software application, provides a graphical user interface for the RouterOS monitoring and configuration. Furthermore, the software allows connections via telnet, FTP, and secure shell (SSH). An application programming interface also can be used for direct access from custom applications for monitoring and management.

One of the most outstanding features of Mikrotik routers is its ability to write our own program to make it more flexible. The packet generation feature is another tool that is able to generate and send raw packets utilizing random or constant port to evaluate performance of system under test (SUT) or device under test (DUT). This traffic generator tool also collects jitter and latency values, counts lost packets, tx/rx rates, and detects Out-of-Order (OOO) packets.

In general, majority of network packets are divided into three parts, which are:

1. *Header.* The header carries information such as length of packet, synchronization, packet number, protocol, destination, and source address.
2. *Payload.* This is also called the data or body of packet. It contains the actual data that the packet is delivering to its destination.
3. *Trailer.* It contains a few bits describing to the destination that the packet is finished. It also may contain some information for error checking.

We generate fake packets using Mikrotik to simulate the attackers $(N_1–N_L)$ in the network depicted in Figure 3.1. To generate this kind of packets, we only manipulate the header of the packets and our algorithm does not care about of contents in payload and trailer of packets generated.

3.2 MULTIROUTER TRAFFIC GRAPHER (MRTG)

The multirouter traffic grapher (MRTG) is a tool, which allows administrators to monitor the traffic load that pass through the network links. MRTG generates HTML pages containing PNG images that

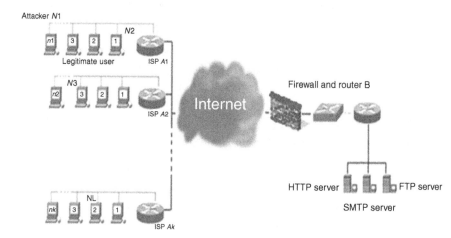

Fig. 3.1. Simulated internet topology.

provide a live visual representation of network traffic. MRTG uses SNMP, which consists of a Perl script to read the traffic counters of routers and a fast C program to collect logs of the traffic data and creates beautiful graphs indicating the traffic on the network connection, which has been monitored. These graphs can be viewed from any modern Web browser.

MRTG creates visual representations of the traffic seen during a detailed daily view, the last seven days, the last five weeks, and the last twelve months. This is possible regarding to the ability of MRTG to keep a log of all the data, which is pulled from the router. MRTG is not limited to monitor traffic, it is also able to monitor any arbitrary SNMP variable. By using an external program, we can even gather the data, which should be monitored via MRTG. Usually, MRTG are using to monitor things such as login sessions, modem availability, system load, and more. MRTG even allows us to accumulate two or more data sources into a single graph.

3.3 BIRTHDAY ATTACK AND BIRTHDAY PARADOX

A birthday attack is a type of cryptographic attack, which exploits the mathematics behind the birthday problem in probability theory. Birthday attack can be used in communication abusage between two or more parties. The attack depends on a fixed degree of permutations (pigeonholes) and the higher likelihood of collisions found between random attack attempts, as described in the birthday paradox/problem.

In probability theory, the birthday paradox or birthday problem considers the probability that some paired people in a set of n randomly chosen of them, will have the same birthday. The mathematics behind this problem led to a well-known cryptographic attack called the birthday attack, which uses this probabilistic model to reduce the complexity of cracking a hash function (McKinney, 1966).

3.4 LEGAL AND ILLEGAL REQUESTS

In this experiment, we modeled the legitimate user and the attacker (illegal user) with several attack strategies of different complexity.

3.4.1 Legitimate Users

Legitimate users are the users who have the legal requests to the accepted destination from the edge routers' point of view (A_1, ..., A_k) and destination firewall routers (B), which are shown in the Figure 3.1. In the other words, legitimate users are end point entities, which make connection requests for services provided by servers.

3.4.2 Illegal Users or Attackers

Attackers are end points whose requests are focused on depleting the target bandwidth or resources. In other words, attackers try to prevent access of legitimate users to the particular server by keeping the number of simultaneous connection as large as possible. Sending session connection requests can be done with stationary rate or totally random.

The most important thing is examining and keeping the number of session connections, which have the same destination address in each local network for specific timeslots under control. This plays an effective role to distinguish and mitigate DDoS attacks.

3.5 TRAFFIC MODELS

To predict and find out how well a flow control scheme works, we must model the behavior of network traffic. A developed model might involve higher-level protocols and characteristics of applications. It is necessary here to clarify and distinguish between "*smooth*" and "*bursty*" traffic.

Predictable and constant load is a result of smooth traffic source, or it might be achieved by changing only on time scales, in which the time scales should be large as compared to the response time of the flow control mechanism. It is easy to interact with this mentioned traffic. The sources can be assigned rates corresponding to fair shares of the bottleneck bandwidth with little risk that some of them will stop sending and lead to underutilized links. Moreover, while bursts in traffic intensity are rare, switches can use a small percentage of available memory.

A good example for source of smooth traffic can be mentioned as fixed-rate compression video and voice traffics. Also, the effect of aggregating of a huge amount of bursty sources may be smooth, especially in a wide area network where the loads from a huge amount of traffic streams are aggregated and the individual sources are uncorrelated and have relatively low bandwidth. On the other hand, bursty traffic lacks any of the predictability of smooth traffic, as observed in some computer communications traffic (Leland, Taqqu, Willinger, & Wilson, 1993). Some types of bursts might be blocked by applications and users. For instance, a user expects to see a page or image just by clicking on a link via World Wide Web browser. The network cannot predict when the clicks will occur, nor should it smooth out the resulting traffic, since doing so would hurt the user's interactive response.

Other sources of bursts result from network protocols that break up transfers into individual packets, remote procedure calls (RPCs), or windows, which are sent at irregular intervals. These bursts do not behave as steady state traffic due to their sporadic properties and typically do not have persistency long enough on a high-speed link to reach steady state over the link round-trip time.

In computer science, a RPC is an inter-process communication that allows a computer program leads to a procedure or a subroutine to implement in another address space (typically on another computer on a shared network) without the programmer explicitly coding the details for this remote interaction. The programmer writes essentially the same code for both local and remote subroutine to the executing program. When the software in question uses object-oriented principles, RPC is called remote invocation or remote method invocation.

Designing control systems for smooth traffic is obviously much easier than designing flow control systems for bursty traffic. One of the most challenging parts, which we face is design effective flow control for bursty traffic in order to support computer communications which are generally bursty.

3.6 ASSUMPTIONS AND CONSIDERATIONS

Consider the network scheme as shown in Figure 3.1. In this scheme, we have assumed that there are k edge routers, which have public IP address (ISP A_1–A_k). We also have assumed that each of these networks has n subscriber in their private local networks. Therefore, there are $n \times k$

separate subscriber as we have also had in our internet network. Moreover, we have considered there are L end users that try to attack specific servers in the same subnet as network B. These users are distributed randomly in the network (N_1-N_L).

Each of $n \times k$ end users is behind NAT. There are several servers such as SMTP, HTTP, and FTP servers in the same subnet as network B, subnet B is behind firewall. In our simulation, router B communicates with each of edge router, which recently have sent request to one of the B subnets. We also introduce some definition as shown in Table 3.1.

3.7 PROBABILITY OF CONCURRENCY REQUEST TO A WEBSITE

According to the recent survey that has been done by Netcraft, an Internet service company based in Bath England, there are around 366,848,493 websites on World Wide Web as of December 2011. It is important for us to know the probability of concurrent outgoing packets, which have same destination address and different source address to set our algorithm threshold level based on that. It means that the algorithm, which is used in edge routers has to determine threshold level for

Table 3.1 Table of Definitions

$\overline{X}(t_m,i)$	Average number of packets or average input traffic load received by a router or firewall during time slot t_m at interface or port i
$\overline{Y}(t_m)$	Average CPU usage
$x(t_m,i)$	Number of packets traffic load received by a router or firewall during time slot t_m at interface or port i
$y(t_m)$	Incident CPU usage at the time slot t_m
β_1	Bandwidth threshold level
β_2	CPU usage threshold level
α	Inertia factor which has to satisfy $0<\alpha<1$ showing the sensitivity of the long-term average behavior to the current traffic variation
$S_{x,in}(t_m,I)$	Deviation of input traffic from the average at time slot t_m
$S_{y,in}(t_m)$	Deviation of incident CPU usage from the average at time slot t_m
$DFA_{x,in}(t_m,i)$	Dimensionless deviation of input bandwidth level from average (indicator of such an attack)
$DFA_{y,in}(t_m)$	Dimensionless deviation of CPU usage from average (indicator of such an attack)

amount of outgoing packets per time slot. The amount higher than this threshold level indicates that the concurrency of some packets with the same destination address and different source address can behave as attack. The birthday attack problem allows us to determine this threshold level. Regarding to the generalized birthday problem, to a group of n people, where $p(n)$ is the probability of at least two of the n people sharing birthday, $p(n)$ would be:

$$p(n) = 1 - \overline{p}(n)$$

where

$$\overline{p}(n) = \frac{n! \binom{365}{n}}{365^n}$$

We can generalize this formula, as n be the amount of Internet users located in a same local area connection, for instance, we had $n = 100$ internet user in our experiment. The total number of different request possibility is the total number of active websites, which has to replace with 365 in the mentioned equation. The probability of concurrency of at least 2 requests to the same destination in a short time slot is also depend on how much the requested website is well-known and famous. For example, there must be difference between the amount of requests to the google.com and a nonfamous personal website. By knowing the statistical average amount of requests, we have to define the time slot. In our experiment, we define 60 s for this time slot, as we knew the average requests to the router B. This means that we do not expect having more than 2 requests with the same destination address during 60 s. There is a trade-off to choose 60 s for this time slot. Any value less than 60 s gives more accurate results but imposes on the routers higher resource usage. For any values more than 60 s, the probability of concurrency of 2 requests is not near zero. Also, this time slot depends strictly to n. If the edge routers find any two concurrent requests in the 60 s time slot, they can assume that something is not normal and the edge routers are going to block the 2 concurrent requests if:

1. The edge routers received a packet from router B containing *"Flag"* tag.
2. The destination addresses of the 2 detected packets are B.

In our simulation in the real network, which we will describe more about, we define three steps to become more accurate to detect the attacks. Each step would take 20 s to monitor all packets in the network.

3.8 DETECTION AND PREVENTION

In general, the mitigation and finding the source of DDoS attacks is a collaboration between edge routers $(A_1, ..., A_k)$ and destination firewall routers B, which has been shown in Figure 3.1. In this collaboration, router B will generate a packet under name of *"Flag"* and send to a group of edge routers one by one when it reaches its specific threshold, which we describe more in detail later. Router B divides all of the edge routers IP address, which has sent request to its network, to j group of IP address in order to raise the speed of finding attackers and mitigation of DDoS attacks. When the edge routers receive this *"Flag"* packet, they run some scripts that we will describe in the next section to find out whether or not the attackers are in their local networks. If they can find any source attacker then they reply the router B with a new packet under name of *"Response to Flag yes"*, which means that we find attacker or attackers and add them to our block list to block their requests. If the edge routers cannot find any suspected source of attacks, they would reply to the router B with a new packet under name of *"Response to Flag no"*.

3.8.1 DDoS Detection Algorithm on Targeted Server

In this part, we will describe how the router B can find out that the server is under DDoS attacks. As we describe about the properties of this kind of attacks, we can write scripts to make the router B sensitive to bandwidth depletion and resource depletion. Regarding to Table 3.1, average number of packets or average input traffic load received by a router during time slot m at port or interface i would be defined as (Chen, Hwang, & Ku, 2007):

$$\bar{X}(t_m, i) = (1 - \alpha) \cdot \bar{X}(t_{m-1}, i) + \alpha \cdot x(t_m, i)$$

And also, the average CPU usage would be derived as:

$$\bar{Y}(t_m) = (1 - \alpha) \cdot \bar{Y}(t_{m-1}) + \alpha \cdot \bar{Y}(t_m, i)$$

where $0 < \alpha < 1$ is an inertia factor showing the susceptibility of the long-term average behavior to the current traffic variation. A higher α implies more dependence on the current variation. Chen et al. (2007) defined $S_{x,in}(t_m,i)$ as the deviation of input traffic from the average at time slot t_m and $S_{y,in}(t_m)$ as the deviation of incident CPU usage from the average at time slot t_m (Chen et al., 2007):

$$S_{x,in}(t_m,i) = \max\left\{0, S_{x,in}(t_{m-1},i) + x(t_m,i) - \bar{X}(t_m,i)\right\}$$

And similarly, we can derive:

$$S_{y,in}(t_m) = \max\left\{0, S_{y,in}(t_{m-1}) + y(t_m) - \bar{Y}(t_m)\right\}$$

The subscript *in* indicates that this is the statistics of the incoming traffic. While a DDoS flooding attack is launched, the cumulative deviation should be noticeably higher than the random fluctuations. Since $S_{x,in}(t_m,i)$ and $S_{y,in}(t_m)$ are sensitive to the changes in the average of the monitored traffic and resource usage respectively, the measurement of the unusual deviation from the historical average can be calculated using the equations mentioned later (Chen et al., 2007). The deviation from average (DFA) is the indicator of such an attack. The incoming traffic DFA is defined below at port i at time t_m (Chen et al., 2007):

$$\mathrm{DFA}_{x,in}(t_m,i) = S_{x,in}(t_m,i)/\bar{X}(t_m,i)$$

And similarly, the CPU usages as the resource depletion at time tm derive as:

$$\mathrm{DFA}_{y,in}(t_m) = S_{y,in}(t_m)/\bar{Y}(t_m)$$

If the $\mathrm{DFA}_{x,in}$ and $\mathrm{DFA}_{y,in}$ exceed a router threshold β_1 and β_2 respectively, the measured traffic surge and resource over usage are considered a suspicious attack. The threshold β_1 and β_2 measure the magnitude of traffic surge and CPU usage over the average traffic and CPU usage value respectively. These parameters are preset based on previous router use experience. In a monitoring window of 100 ms to 1 s, a normal superflow is rather smooth due to statistical multiplexing of all independent flows heading for the same destination (Jiang & Dovrolis, 2005).

We expect a small deviation rate far below β_1 and β_2 if there is no DDoS attack. A superflow contains all packets destined for the same network domain from all possible source Internet protocol (IP) addresses and applies various protocols such as transmission control protocol (TCP) or user datagram protocol (UDP), etc. In general, we assume the work range for parameter β_1 and β_2 in the range of $2 \leq \beta \leq 5$. In order to achieve best results, we run our algorithm with two different values of router threshold setting β and compared results. Chen et al. (2007) achieved an optimal router threshold setting $\beta \geq 3.5$ with an inertia ratio $\alpha = 0.1$. We choose $\beta = 3$ with an inertia ratio $\alpha = 0.1$ and $\beta = 4$ with an inertia ratio $\alpha = 0.1$ to make a trade off and find an appropriate router threshold setting.

Regarding to the definition of β, choosing any value below 2 makes the algorithm very sensitive to any small fluctuation of input traffic and CPU usage. Figure 3.2 shows the difference between choosing the β_1 value. In this graph, the arrow shows the attack traffic. By setting any value below 3 for β_1, false-positive errors would be increased by detecting legal traffic request as attacks. On the other hand, by setting any value more than 5 for β_1, false-negative errors would increase as the algorithm allows more attacks pass through the edge routers.

When router B reaches the threshold level of β_1 and β_2 first the server put every 100-edge router's IP address in different j block or group. Next, the server generates a *Flag* packet to send to the first group, which

"Daily" graph (5 min average)

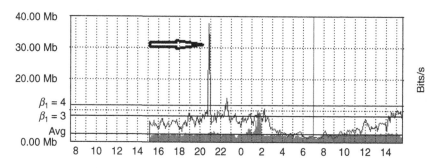

Max *In*: 9.81Mb; Average *In*: 2.19Mb; Current *In*: 1.37Mb;
Max *Out*: 37.98Mb; Average *Out*: 5.38Mb; Current *Out*: 7.76Mb;

Fig. 3.2. Effect of parameter β_1.

is $j = 1$ and wait for the response. After that, the router B has to wait due to running process in the edge routers for making any response. Once the waiting time has elapsed, the router B begins to receive response packet from the first group of edge routers, which are either marked as "*Response to Flag yes*" or "*Response to Flag no*". In general, the whole process, which is used in both targeted router B and edge routers $A_1, \ldots, A_k$ is illustrated briefly as:

1. Router B Calculate $\overline{X}(t_m, i)$, $\overline{Y}(t_m)$, $S_{x,in}(t_m, i)$, $S_{y,in}(t_m)$, $DFA_{x,in}(t_m, i)$, and $DFA_{y,in}(t_m)$.
2. If $DFA_{x,in}(t_m, i) > \beta_1 \rightarrow$ go to 3 else go to 1.
3. If $DFA_{y,in}(t_m) > \beta_2 \rightarrow$ go to 4 else go to 1.
4. Send a *Flag* request from router B to a group or all of edge routers to ask them to check their networks for attacks.
5. Edge routers $A_1, \ldots, A_k$ receive the *Flag* request and commence to monitor all outgoing packets.
6. If find at least 2 packet with the destination address of B and different source address in period of 60 s, which is divided into three steps of 20 s $\rightarrow$ go to 7 else go to 8.
7. Block or redirect this packet and send a reply packet to router B under the name of "*Respond to Flag Yes*" $\rightarrow$ go to 1.
8. Send a reply packet to router B under the name of "*Response to Flag No.*"
9. Go to 1.

The flowchart in Figure 3.3 states the step-by-step sections of implementing scripts in the router B as the destination of DDoS attacks.

3.8.2 DDoS Detection Algorithm on Edge Routers

As shown in Figure 3.1, edge router is operating at the edge of an multiprotocol label-switching network that provides entry points into enterprise or service provider core networks. Edge routers also provide connections into carrier and service provider networks. We have to underline the importance of the role of these routers to help to identify and mitigate the DDoS attacks. In other words, these routers act as a gate of entrance for the local and public packets to the outside and inside a local area respectively. These gates also have access to the detail of packets, which flows through. With utilizing these gates, we can identify and mitigate DDoS attacks.

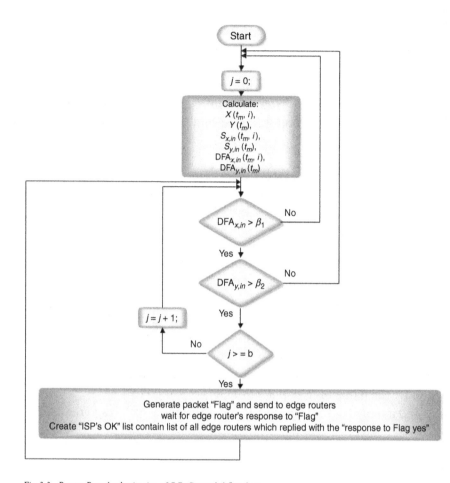

Fig. 3.3. Router B as the destination of DDoS attacks' flowchart.

The router B generate packets and send to the destination address listed in group j and particular port of x, which is known by routers A_1 to A_k. In this step, router B is waiting for reply from each edge routers, which are addressed in each block respectively. In edge routers, we wrote a script as a scheduler to check, if there is an incoming packet with destination port number of x to activate another script, which is in charge of detecting attackers in its private network. The algorithm, which is used in the Mikrotik firewall edge routers, has been shown in Figure 3.4.

If the edge routers detect any attacks, reply to B with a packet, which has destination port number of x as *'Response to Flag yes'* otherwise

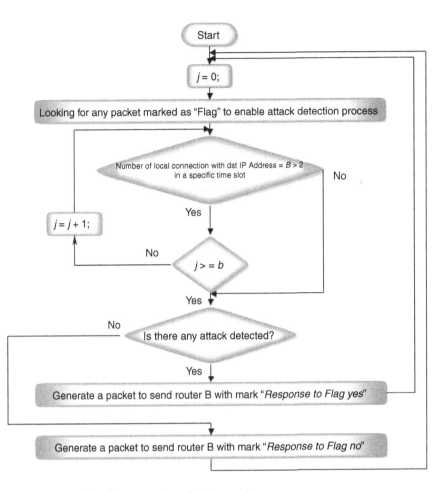

Fig. 3.4. Flowchart of the edge routers scripts to distinguish attackers.

reply with destination port number of *y* as *'Response to Flag no'*, which are known in the firewall router *B*. Router *B* get packet from all edge routers in the first block or some of them. If current received bandwidth and current CPU usage become below the threshold level then save the first block in the list of suspected source of attacks for its own records, the same procedure is repeated to the second block of edge routers IP address list. These records can be also exchanged with other firewall servers.

CHAPTER 4

Results and Discussions

Maintaining Internet users always online requires service providers' strong security response to mitigate distributed denial of service (DDoS) attacks. Why is early detection so important? Time is literally money when mitigating a DDoS attack, data breach, or other type of cyber-attack. Faster DDoS detection allows DDoS mitigation to begin more quickly, which saves money and reduces the impact of damaged brand reputation, lost customers, and declining stock prices. Most IT organizations do not have the specialist skill set to perform around-the-clock DDoS monitoring and DDoS detection. A specialized DDoS detection service as part of a larger cloud security service can provide advanced visibility into global Internet traffic and traffic at your website or data center. Dedicated technicians in a specialized Security Operations Center (SOC) can monitor customers' networks 24/7 for early DDoS detection of malicious application-layer and network-layer traffic. Seasoned web security experts on the DDoS detection provider's security team can also serve as web security consultants who ensure that web applications and network systems are always up-to-date and protected against emerging threats.

In our topology that has been shown in Figure 3.1, we use Mikrotik routers as end clients to generate random UDP and TCP packets to destination *B*. Mikrotik Linux-based firewall server gave us simulated behavior of attacks in large networks as well. The MRTG graph which is shown in Figure 4.1 shows the daily incoming and outgoing traffic of router *B* which takes 5 min average. Figure 4.2 shows the daily CPU usage of 5 min average under DDoS attack simulation and Figure 4.3 shows the daily CPU usage of 5 min average without facing any attack. We generated random traffic to simulate DDoS attacks.

To generate random packets that behave like attackers, we have used Mikrotik router. By using packet generator, which is a Mikrotik tool, we generated fake traffic to the destination router *B*. The content of packet

"Daily" graph (5 min average)

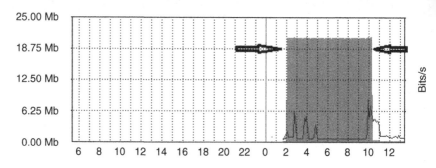

Max In: 21.19Mb; Average In: 14.27Mb; Current In: 37.07Kb;
Max Out: 8.51Mb; Average Out: 1.33Mb; Current Out: 763.04Kb;

Fig. 4.1. MRTG daily incoming (grey area) traffic into LAN of edge router before launching our algorithm.

"Daily" graph (5 min average)

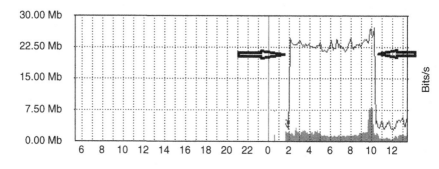

Max In: 7.89Mb; Average In: 1.59Mb; Current In: 1.15Mb;
Max Out: 27.35Mb; Average Out: 17.11Mb; Current Out: 5.18Mb;

Fig. 4.2. MRTG daily outgoing (single line) traffic via WAN of edge router before launching our algorithm.

does not matter because the algorithm only analyze the packet headers. Source of packet, destination of packet, and number of packet sent by every individual user are three main factors that must be taken into consideration to behave as attack. After generating random traffic, we run the algorithm presented in Chapter 3 with router threshold setting of

- $\beta = 3$ with an inertia ratio $\alpha = 0.1$.
- $\beta = 4$ with an inertia ratio $\alpha = 0.1$.

"Daily" graph (5 min average)

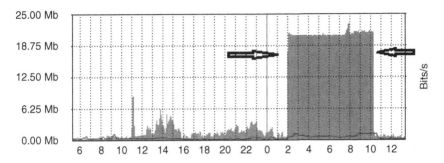

Max In: 23.16Mb; Average In: 6.37Mb; Current In: 286.45Kb;
Max Out: 1.18Mb; Average Out: 305.76Kb; Current Out: 27.51Kb;

Fig. 4.3. MRTG daily incoming (grey area) traffic into LAN of router B under attack before launching our algorithm.

We could detect simulated attacks and block the attackers very well. Figure 4.2 shows the generated attack from the edge router's LAN. The destination of these packets is router *B*. The edge router and router *B* refers to those seen in Figure 3.1. For further clarity, the different interfaces on these routers are shown in Figure 4.4.

Edge routers consider the whole traffic as good traffic and route them to their destination address through its WAN interface that is shown in Figure 4.2. In other words, the edge routers do not care about where the packets are destined or whether the packets are potentially DDoS traffic. Figure 4.3 shows that the same amount of traffic that has been passed through the edge routers now act as attack traffic for router *B* LAN interface to deplete its bandwidth. This huge amount of request to the router *B* not only deplete a considerably percentage of available link bandwidth, but also can lead to deplete the resource in router *B* as shown in Figure 4.5. Figure 4.6 illustrates what happens when an attack is launched while running our algorithm. This figure shows the incoming generated attack packets in the LAN of edge router. For a short period, edge router do not care about the traffic and let the packets go through. Router *B* finds out about the attack due to excessive use of the CPU and deviation of incoming average traffic from threshold level as shown in Figure 4.7 and Figure 4.8. At this time, router *B* send request to edge router as described in detail in Chapter 3 to enable scripts to commence the process of detecting and mitigating attack. After a while,

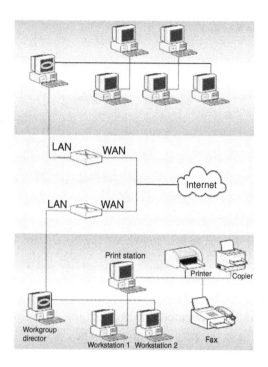

Fig. 4.4.   General WAN/LAN diagram.

"Daily" graph (5 min average)

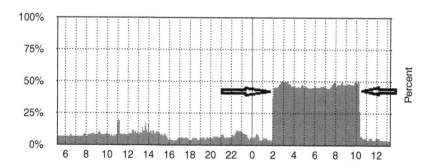

Max: 50%; Average: 17%; Current: 3%;

Fig. 4.5.   MRTG daily CPU usage of router B under attack before launching our algorithm.

"Daily" graph (5 min average)

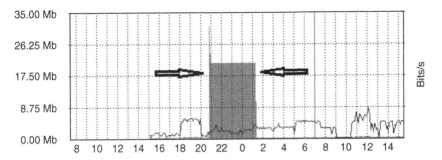

Max In: 31.53Mb; Average In: 4.02Mb; Current In: 119.36Kb;
Max Out: 8.68Mb; Average Out: 2.85Mb; Current Out: 4.61Mb;

Fig. 4.6. MRTG daily incoming (grey area) traffic into LAN of edge router during running our algorithm.

"Daily" graph (5 min average)

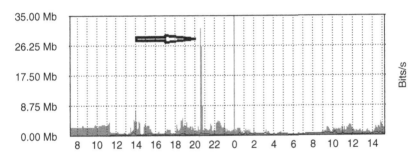

Max In: 31.43Mb; Average In: 3.15Mb; Current In: 1.67Mb;
Max Out: 1.18Mb; Average Out: 216.93Kb; Current Out: 71.18Kb;

Fig. 4.7. MRTG daily incoming (grey area) traffic into LAN of router B during running our algorithm.

edge router detects and mitigates the attack and block all sources of attack. The spikes in Figure 4.9, Figure 4.7, and Figure 4.8 which are shown with arrows indicate the short period of time that take to detect and mitigate the whole attack process by collaborating with the two most important elements of network that are edge routers and destined routers.

"Daily" graph (5 min average)

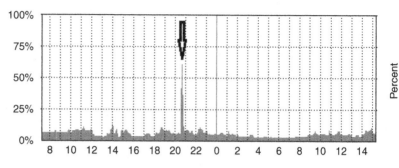

Max: 61%; Average: 9%; Current: 5%;

Fig. 4.8. MRTG daily CPU usage of router B during running our algorithm.

"Daily" graph (5 min average)

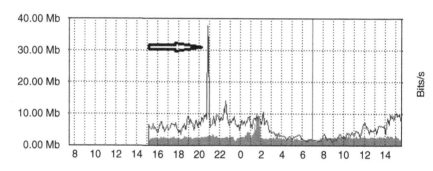

Max In: 9.81Mb; Average In: 2.19Mb; Current In: 1.37Mb;
Max Out: 37.98Mb; Average Out: 5.38Mb; Current Out: 7.76Mb;

Fig. 4.9. MRTG daily outgoing (single line) traffic via WAN of edge router during running our algorithm.

Time taken for the whole process is shown in Tables 4.1 and 4.2 for ten independent experiments. Results for false-negative and false-positive error for different router setting for the two different parameters of β and α in ten independent experiments are shown in Tables 4.3 and 4.4.

Our algorithm is able to detect almost all illegal requests in the form of UDP and TCP regardless of their source port, destination port, and IP address.

Table 4.1 Time Taken for the Whole Detection and Mitigation Process for Router Setting of $\beta = 4$ and $\alpha = 0.1$

No. of Experiments	1	2	3	4	5	6	7	8	9	10
Time elapsed for the whole process in seconds	123	127	121	129	124	127	124	129	128	126

Table 4.2 Time Taken for the Whole Detection and Mitigation Process for Router Setting of $\beta = 3$ and $\alpha = 0.1$

No. of Experiments	1	2	3	4	5	6	7	8	9	10
Time elapsed for the whole process in second	113	120	114	120	116	120	115	121	119	117

Table 4.3 Results for False-Negative and False-Positive Error for Router Setting of $\beta = 4$ and $\alpha = 0.1$

No. of Experiments	1	2	3	4	5	6	7	8	9	10
No. of attacks not detected	1	2	1	0	0	2	1	1	0	2
Probability ($P-$) of not detecting attack traffic (%)	6.66	13.3	6.66	0	0	13.3	6.66	6.66	0	13.3
No. of points erroneously detected as attacks	0	1	0	0	0	0	0	1	0	0
No. of points detected as attacks in total	14	12	14	15	15	13	14	13	15	13
Probability ($P+$) of erroneously detecting attack (%)	0	6.66	0	0	0	0	0	6.66	0	0

Table 4.4 Results for False-Negative and False-Positive Error for Router Setting of $\beta = 3$ and $\alpha = 0.1$

No. of Experiments	1	2	3	4	5	6	7	8	9	10
No. of attacks not detected	0	1	0	0	0	0	0	1	0	0
Probability ($P-$) of not detecting attack traffic (%)	0	6.66	0	0	0	0	0	6.66	0	0
No. of points erroneously detected as attacks	3	1	4	2	1	4	4	1	3	3
No. of points detected as attacks	12	13	11	13	14	11	11	13	12	12
Probability ($P+$) of erroneously detecting attack (%)	25	6.66	36.4	15.4	6.6	26.6	26.6	6.66	20.0	20.0

4.1 TIME INVESTIGATION IN ATTACK DETECTION

Tables 4.1 and 4.2 show the time, which was taken to detect source IP address of attackers and mitigating the DDoS attacks in the whole process containing detection and mitigation as well for different router threshold setting.

The average time taken in Table 4.1 is 125.8 s whereas this average time in Table 4.2 is 117.5 s, which means that using lower threshold level β will lead to faster detection of the attacks.

4.2 FALSE-POSITIVE AND FALSE-NEGATIVE ERROR

A false positive error, commonly called a "false alarm" is a result, which indicates a given condition has been fulfilled, when it actually has not been fulfilled. On the other hand, a false negative error is where a test result indicates that a condition was failed, while it actually was successful.

We define the probability $(P-)$ of not detecting the attack traffic (ie, the probability of the false-negative errors) and the probability $(P+)$ of erroneously detecting an attack (ie, the probability of false-positive errors), which are calculated from the following equations:

$$P- = \frac{\text{Number of attacks not detected}}{\text{Total number of attacks}}$$

$$P+ = \frac{\text{Number of points errorneously detected as attacks}}{\text{Total number of attacks}}$$

We generated fake packets that behave as attacks from 15 different locations in the network by 15 independent Mikrotik devices. Hence, the number of attackers in our experiment is 15 (N_1–N_{15} in Figure 3.1) attacks in total.

As shown in Tables 4.3 and 4.4, the results of repeating the same experiment for 10 rounds in the real network imply on the accuracy and trustworthiness of the proposed algorithm. Note that in Table 4.1, Table 4.3, and Table 4.5, all the measurements are considered regarding to router threshold setting of $\beta = 4$ with an inertia ratio $\alpha = 0.1$. All

Table 4.5 Detection Rate R_d for $\beta = 4$ and $\alpha = 0.1$

No. of Experiments	1	2	3	4	5	6	7	8	9	10
Detection ratio (R_d) (%)	93.3	80	93.3	100	100	86.6	93.3	86.6	100	86.6

Table 4.6 Detection Rate R_d for $\beta = 3$ and $\alpha = 0.1$

No. of Experiments	1	2	3	4	5	6	7	8	9	10
Detection ratio (R_d) (%)	80	86.6	73.3	86.6	93.3	73.3	73.3	86.6	80	80

the measurements in Table 4.2, Table 4.4, and Table 4.6 are for router threshold setting of $\beta = 3$ with an inertia ratio $\alpha = 0.1$.

4.3 MEASURING THE PERFORMANCE METRICS

The performance of our detection scheme is evaluated with three following metrics: detection rate, false-positive alarms ($P+$), and system overhead. All the metrics are measured under different DDoS attacks using TCP, UDP, and ICMP. The detection rate R_d of DDoS attacks is defined by the following ratio (Chen, Hwang, & Ku, 2007):

$$R_d = \frac{a}{n}$$

where a is the number of DDoS attacks detected in the simulation experiments, and n is the total number of attacks generated by the Mikrotik routers during each experiment which is 15. Tables 4.5 and 4.6 show the measured detection ratio over the whole ten independent experiments.

4.4 TRADE OFF

By taking a glance view on the Tables 4.1 and 4.2, we can see that the number of attacks not detected in Table 4.1 is more than Table 4.2 due to the differences between the two parameters of β and α. The most important thing to achieve the best results is to keep false-positive alarm rate minimum. If we look carefully on the result of these two tables, we can see that the number of points erroneously detected as attacks in the Table 4.1 are less than the Table 4.2 by far. Figure 4.10 compares the detection rates R_d between the results achieved by these two tables.

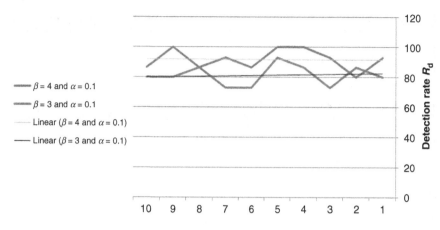

Fig. 4.10. Detection ratio R$_d$ with different router threshold levels (α and β).

The linear lines in Figure 4.10 are trend line for two different threshold levels. These two trend lines show a trade off on average detection ratio R_d for different router threshold settings. According to the achieved results, we can conclude that using our algorithm with the optimal router threshold setting $\beta \geq 3.5$ with an inertia ratio $\alpha = 0.1$ give us the best results.

4.5 SUMMARY

We must underline the importance of two following parameters that plays the most challenging role to select suitable algorithm to identify and mitigate DDoS attacks:

1. Achieve higher detection rate
2. Achieve low detection time

To achieve higher detection rate in the lowest time, we also have to consider and keep the two main factors of false-positive and false-negative errors optimum. In this work, we tried to satisfy these factors as well.

CHAPTER 5

Conclusions and Recommendations

5.1 CONCLUSIONS

To avoid any widespread damage to the victims system, DDoS attacks have to be detected at their early launching stage. We have developed a defense and distributed detection mechanism to protect servers and websites against distributed denial of service (DDoS) attacks. We proposed a mechanism based on collaboration among the two most important gateways of attacks which are edge routers and victims firewall routers. Such a network is an effective way to enhance the attacks detection rate, provide attack alerts, and protect legitimate traffic.

We proposed a new detection method to increase the attacks detection accuracy, which samples the current incoming traffic and CPU usage of the destination target of attack and calculates the difference from average. It then compares it to the predefined parameter β to distinguish whether or not the attacks happened. The threshold β measures the magnitude of traffic surge over the average traffic value or the magnitude of CPU surge over the average CPU value. If the firewall senses any attacks, it will send a request to any source edge routers to collaborate with them in order to detect and mitigate attacks.

Based on this result achieved, we introduced a new attack detection method. Through real experiments, we have shown that our method can detect and block attack packets quickly. We have also shown the effects of attacker-side defense and the effectiveness of our method.

5.2 RECOMMENDATIONS

In the large networks, for a faster response to detect and mitigate attacks, we recommend the categorization of edge router in j groups. It would be more efficient to send a collaborative Flag request to the whole edge routers simultaneously. This categorization can be done in the first step by dividing the number of available edge routers to two independent groups and each also divide by two. The number of division iterations would depend on how fast we expect to find attacks and the number of available edge routers.

REFERENCES

Abadi, M., et al. (2005). Moderately hard, memory-bound functions. *ACM Transactions on Internet Technologies, 5*, 299–327.

Aljifri, H. (2003). IP traceback: a new denial-of-service deterrent. *IEEE Security and Privacy*, 24–31.

Aura, T., et al. (2001). DOS-resistant authentication with client puzzles. In B. Christianson et al. (Ed.), *Security Protocols* (pp. 170–177). (2133). Berlin/Heidelberg: Springer.

Bellovin, S., Schiller, J., & Kaufman, C. (2003). Security mechanism for the internet. *IETF RFC, 3631*, 1–101.

Blazek, R., et al. (2001). A novel approach to detection of DoS attacks via adaptive sequential and batch-sequential change-point detection methods. *Proceedings on IEEE workshop information assurance and security*.

Carl, G., Kesidis, G., Brooks, R., & Rai, S. (2006). Denial-of-service attack detection techniques. *IEEE Internet Computing, 10*(1), 82–89.

Chen, Y., & Hwang, K. (2006a). Collaborative change detection of DDoS attacks on community and ISP networks. *Proceedings of IEEE international symposium on collaborative technologies and systems* (CTS'06).

Chen, Y., & Hwang, K. (2006b). Collaborative detection and filtering of shrew DDoS attacks using spectral analysis. *Journal of Parallel and Distributed Computing, 66*(9), 1137–1151 (special issue on security in grids and distributed systems).

Chen, S., & Song, Q. (2005). Perimeter-based defense against high bandwidth DDoS attacks. *IEEE Transactions on Parallel and Distributed Systems, 16*(6), 526–537.

Chen, Y., Hwang, K., & Ku, W. S. (2007). Collaborative detection of DDoS attacks over multiple network domains. *IEEE Transactions on Parallel and Distributed Systems, 18*(12), 1649–1662.

Culberson, J. C., & Schaeffer, J. (1996). Searching with pattern databases. *Proceedings of the eleventh biennial conference of the Canadian society for computational studies of intelligence on advances in artificial intelligence*.

Dean, D., & Stubblefield, A. (2001). Using client puzzles to protect TLS. *Proceedings of the 10th conference on USENIX Security Symposium*. Vol. 10, Washington, D.C.

Doshi, S., Monrose, F., & Rubin, A. D. (2006). Efficient memory bound puzzles using pattern databases. *ACNS, 3989*, 98–113.

Feng, W., & Kaiser, E. (2007a). mod_kaPoW: mitigating DoS with transparent proof-of-work. *Proceedings of the 2007 ACM CoNEXT conference*. New York.

Feng, W., & Kaiser, E. (2007b). The case for public work. *Proceedings of global internet 2007*.

Feng, W., & Kaiser, E. (2010). kaPoW webmail: effective disincentives against spam. *CEAS 2010*.

Feng, W., et al. (2005). Design and implementation of network puzzles. *Proceedings of IEEE INFOCOM 2005, twenty fourth annual joint conference of the IEEE computer and communication societies* (pp. 2372–2382).

Ferguson, P., & Senie, D. (2000). Network ingress filtering: defeating denial of service attacks which employ IP source address spoofing. RFC 2827.

Fraser, N.A., Kelly, D.J., Raines, R.A., Baldwin, R.O., & Mullins, B.E. (2007). Using client puzzles to mitigate distributed denial of service attacks in the tor anonymous routing environment. *Proceedings of ICC 2007*.

Fullmer, M., & Romig, S. (2000). The OSU flowtools package and cisco netflow logs. *Proceedings of the 2000 USENIX LISA Conference*. New Orleans, LA.

Gao, Y., et al. (2010). Efficient trapdoor-based client puzzle against DoS attacks. In S. C. H. C. H. Huang et al. (Ed.), *Network Security* (pp. 229–249). US: Springer.

Gil, T., & Poletto, M. (2001). MULTOPS: a data-structure for bandwidth attack detection. *Proceedings of the tenth USENIX Security Symposium*.

Groza, B., Petrica, D. (2006). On chained cryptographic puzzles. *The third Romanian-Hungarian joint symposium on applied computational intelligence (SACI)*, Timisoara, Romania.

Houle et al., (2001). Trends in denial of service attack technology. www.cert.org/archive/pdf/.

Hussain, A., Heidemann, J., & Papadopoulos, C. (2006). Identification of repeated denial of service attacks. *Proceedings of INFOCOM '06*.

Hwang, K., Cai, M., Chen, Y., & Qin, M. (2007). Hybrid intrusion detection with weighted signature generation over anomalous internet episodes. *IEEE Transactions on Dependable and Secure Computing*, 4(1), 41–55.

Ioannidis, J., & Bellovin, S.M. (2002). Implementing pushback: router-based defense against DDoS attacks. *Proceedings of network and distributed system security symposium (NDSS '02)*.

Jeckmans, A. J. P. (2009). Practical client puzzle from repeated squaring. *Centre for Telematics and Information Technology*. University of Twente, Netherlands.

Jiang, H., & Dovrolis, C. (2005). Why is the internet traffic bursty in short time scales. *Proceedings of the ACM SIGMETRICS '05*.

Jin, C., Wang, H., & Shin, K.G. (2003). Hop-count filtering: an effective defense against spoofed traffic. *Proceedings of ACM conference on computer and communication security (CCS '03)*.

Juels, A., & Brainard, J. G. (1999). Client puzzles: a cryptographic countermeasure against connection depletion attacks. *Proceedings of the network and distributed system security symposium* (pp. 151–165).

Kandula, S., Katabi, D., Jacob, M., & Berger, A. (2005). Botz-4-sale: surviving organized DDoS attacks that mimic flash crowds. *Proceedings of the second symposium on networked systems design and implementation* (NSDI'05).

Karame, G., & Čapkun, S. (2010). Low-cost client puzzles based on modular exponentiation. In D. Gritzalis et al. (Ed.), *Computer Security – ESORICS 2010* (pp. 679–697). (vol. 6345). Berlin/ Heidelberg: Springer.

Kent, S., & Atkinson, R. (1998). Security architecture for the internet protocol. *IETF RFC, 2401*, 1–101.

Keromytis, A. D., Misra, V., & Rubenstein, D. (2004). SOS: an architecture for mitigating DDoS attacks. *IEEE Journal of Selected Areas in Communication*, 22(1), 176–188.

Kim, Y., Jo, J.Y., & Merat, F. (2003). Defeating distributed denial-of-service attack with deterministic bit marking. *Proceedings of IEEE GLOBECOM*.

Kim, Y., Jo, J.Y., Chao, H.J., & Merat, F. (2003). High-speed router filter for blocking tcp flooding under distributed denial-of-service attack. *Proceedings of IEEE international performance, computing, and communication conference*.

Kim, Y., Lau, W.C., Chuah, M.C., & Chao, H.J. (2004). PacketScore: statistics-based overload control against distributed denial of service attacks. *Proceedings of IEEE INFOCOM '04*.

Kuzmanovic, A. & Knightly, E.W. (2003). Low-rate TCP-targeted denial of service attacks (The Shrew vs. the Mice and Elephants). *Proceedings of ACM SIGCOMM 2003*.

Leland, W.E., Taqqu, M.S., Willinger, W., & Wilson, D.V. (1993). On the self-similar nature of ethernet traffic. *Proceedings of the ACM SIGCOMM 1993 symposium on communications architectures, protocols, and applications* (pp. 183–193).

Lenstra, A. K., et al. (1982). Factoring polynomials with rational coefficients. *Mathematische Annalen, 261*, 515–534.

Li, Q., Chang, E., & Chan, M. (2005). On the effectiveness of DDoS attacks on statistical filtering. *Proceedings of INFOCOM'05.*

Loyd, S., & Gardner, M. (1959). *Mathematical puzzles.* Courier Corporation, USA.

McKinney, E. H. (1966). Generalized birthday problem. *American Mathematical Monthly, 73,* 385–387.

Merkle, R. C. (1978). Secure communications over insecure channels. *Communications of the ACM, 21,* 294–299.

Mirkovic, J., & Reiher, P. (2004). A taxonomy of DDoS attack and DDoS defense mechanisms. *ACM SIGCOMM Computer Communication Review, 34*(2), 39–53.

Mirkovic, J., & Reiher, P. (2005). D-WARD: a source-end defense against flooding DoS attacks. *IEEE Transactions on Dependable and Secure Computing, 2*(3), 216–232.

Mirkovic, J., Prier, G., & Reiher, P. (2002). Attacking DDoS at the Source. *Proceedings of the tenth IEEE International Conference on Network Protocols.*

Moore, D., Voelker, G., & Savage, S. (2001). Inferring internet denial-of-service activity. Proceedings of the tenth USENIX security symposium.

Ning, P., Jajodia, S., & Wang, X. S. (2001). Abstraction-based intrusion detection in distributed environment. *ACM Transactions on Information and System Security, 4*(4), 407–452.

Papadopoulos, C., Lindell, R., Mehringer, J., Hussain, A., & Govindan, R. (2003). COSSACK: coordinated suppression of simultaneous attacks. *Proceedings of the third DARPA information survivability conference and exposition (DISCEX-III '03)* (pp. 2–13).

Park, K., & Lee, H. (2001). On the effectiveness of route-based packet filtering for distributed DoS attack prevention in power-law internets. *ACM SIGCOMM Computer Communication Review, 31*(4).

Peng, T., Leckie, C., & Ramamohanarao, K. (2003). Detecting distributed denial of service attacks by sharing distributed beliefs. *Proceedings of the eighth Australasian conference information security and privacy (ACISP '03).*

Ranjan, S., Swaminathan, R., Uysal, M., & Knightly, E. (2006). DDoS resilient scheduling to counter application layer attacks under imperfect detection, *Proceedings of IEEE INFOCOM.*

Rivest, R. L., et al. (1996). *Time-lock puzzles and timed-release crypto.* Cambridge, Mass.: MIT Laboratory for Computer Science.

Ryutov, T., Zhou, L., Neuman, C., Leithead, T., & Seamons, K.E. (2005). Adaptive trust negotiation and access control. *Proceedings of ACM symposium access control models and technologies (SACMAT'05).*

Specht, S.M. & Lee, R.B. (2004). Distributed denial of service: taxonomies of attacks, tools, and countermeasures. *Proceedings of the seventeenth international conference on parallel and distributed computing systems, international workshop on security in parallel and distributed systems* (pp. 543–550).

Tritilanunt, S. (2010). Performance evaluation of non-parallelizable client puzzles for defeating DoS attacks in authentication protocols. In S. Foresti, & S. Jajodia (Eds.), *Data and applications security and privacy XXIV* (pp. 358–365). (vol. 6166). Berlin/Heidelberg: Springer.

Tritilanunt, S., et al. (2007). Toward non-parallelizable client puzzles. *Proceedings of the sixth international conference on cryptology and network security,* Singapore.

Walfish, M., Vutukuru, M., Balakrishnan, H., Karger, D., & Shenker, S. (2006). DDoS defense by offense. *Proceedings of the ACM SIGCOMM'06.*

62   References

Wang, H., Zhang, D., & Shin, K. (2004). Change-point monitoring for the detection of DoS attacks. *IEEE Transactions on Dependable and Secure Computing, 1,* 193–208, http://www.mikrotik.com/.

Wang, X., Chellappan, S., Boyer, P., & Xuan, D. (2006). On the effectiveness of secure overlay forwarding systems under intelligent distributed DoS attacks. *IEEE Transactions on Parallel and Distributed Systems, 17*(7), 619–632.

Waters, B., et al. (2004). New client puzzle outsourcing techniques for DoS resistance. *Proceedings of the eleventh ACM conference on computer and communications security,* Washington DC, USA.

Wikipedia, http://en.wikipedia.org/wiki/Birthday_attack.

Xu, Y., & Guérin, R. (2005). On the robustness of router-based denial of-service (DoS) defense systems. *ACM SIGCOMM Computer Communication Review, 35*(3), 47–60.

Yaar, A., & Song, D. (2004). SIFF: a stateless internet flow filter to mitigate DDoS flooding attacks. *Proceedings of 2004 IEEE symposium of security and privacy.*

Yu, J., Fang, C., Lu, L., & Li, Z. (2009). A lightweight mechanism to mitigate application layer DDOS attacks. In *Scalable information systems* (pp. 175–191). (vol. 18). Berlin: Springer.

Printed in the United States
By Bookmasters